How We Use Paper

Chris Oxlade

Chicago, Illinois

Published by Raintree,
A division of Reed Elsevier, Inc.
Chicago, IL

For information, address the publisher:
Raintree
100 N. LaSalle, Suite 1200
Chicago, IL 60602

Originated by Ambassador Litho
Printed and bound in China by
South China Printing Company

09 08 07 06 05
10 9 8 7 6 5 4 3 2 1

Library of Congress Cataloging-in-Publication Data

Oxlade, Chris.
How we use paper / Chris Oxlade.
p. cm. -- (Using materials)
Includes bibliographical references and index.
ISBN 1-4109-0603-5 (library binding-hardcover)
-- ISBN 1-4109-0894-1 (pbk.)
1. Paper--Juvenile literature. I. Title. II. Series.
TS1105.5.O95 2004
676--dc22

2004002914

Acknowledgments

The publisher would like to thank the following for permission to reproduce photographs: pp. 4 (Andrew Syed), 7 (Geoff Tompkinson), 8 (Ben Johnson), 14 (George Bernard) Science Photo Library; pp. 5 (Harcourt Index), 6 (Harcourt Index), 11 (Ron Watts), 19 (Kelly-Mooney Photography) Corbis; p. 9 (Jack Culbertson) Powerstock; p. 10 (H. Rogers) Trip; pp. 12 (Photodisc/Harcourt Index), 18 (Photodisc), 22 (Foodpix), 23 (Imagebank), 27 (Taxi/Arthur Tilley) Getty Images; pp. 13, 16, 17, 21, 25 Alamy Images; pp. 15 (Picture Tasmania Photolibrary/Steve Lovegrove), 26 (Gillian Gunner Photography) photographersdirect.com; pp. 20, 24 Tudor Photography; p. 28 Ecoscene; p. 29 (AA/Patti Murray) OSF.

Cover photograph of rolls of manufactured paper ready for collection reproduced with permission of Alamy.

Every effort has been made to contact copyright holders of any material reproduced in this book. Any omissions will be rectified in subsequent printings if notice is given to the publishers.

Contents

Some words appear in bold, **like this.** You can find out what they mean by looking in the glossary.

Paper and Its Properties

All the things we use are made from **materials.** Paper is a material. We use paper for many different jobs. Its main use is for printing and writing. Newspapers, magazines, and books are all printed on paper. We use a lot of paper in computer printers. We also make tissues, decorations, and objects such as cups and packaging from paper.

This is how paper appears when looked at through a **microscope.**

Paper is good for writing on because ink sticks to its surface.

The **properties** of a material tell us what it is like. Paper is bendable and easy to fold. When in thin sheets, paper is also weak and easy to tear. Cardboard is strong and hard to tear, however. Paper is not **waterproof,** and it burns very easily. It can be many different colors.

Do not use it!

The different properties of materials make them useful for some jobs. The properties also make them not good for other jobs. For example, paper is weak. For this reason, it would be hard to make a chair from paper that would be strong enough to sit on.

What Is Paper Made From

Paper is not a **natural material.** But it is made from natural materials. We make it from parts of plants and trees called **fibers**. Most fibers for papermaking come from the wood of trees.

The first step in making paper is to make **pulp.** Pulp is a mixture of water and fibers. **Mechanical pulp** is made by crushing wood. **Chemical pulp** is made by soaking the wood in **chemicals** that make it fall apart. **Rag pulp** is made from the fibers of cotton and linen plants.

A factory where wood is made into paper is called a paper mill.

This man is checking the quality of wood pulp in a paper mill.

The pulp is mixed with water to make a thick liquid. It is then poured onto a wire **mesh.** Water drains through the gaps in the mesh. But the fibers are trapped. The layer of fibers is rolled to squeeze out any water that is left. A thin sheet of damp paper is left. The paper is then dried.

Paper in history

The word paper *comes from the word* papyrus. *The ancient Egyptians wrote on a material made from the leaves of the papyrus reed. Paper made from tree fibers was invented in China about 2,000 years ago.*

Paper for Writing and Books

We use paper for writing letters and for making books. Paper is good for these jobs for a few reasons. First, it is easy to cut and fold. It is also lightweight, and ink sticks to it well. The main types of paper used for these jobs are book paper and bond paper.

Book paper whizzes through a printing press at high speed.

Book paper is usually made from **chemical pulp,** but it is sometimes made from **mechanical pulp.** Book paper is not as strong or thick as bond paper, which is used for writing. **Pigments** are added to the paper to make it white.

These pages are about 0.004 inch (0.1 mm) thick.

Bond paper is used for writing and drawing. It is thick, smooth, and strong. Writing and printing paper must have smooth surfaces that ink will stick to. The paper must also stop the ink from soaking in and getting **blotchy.** Paper is coated with a substance called size to make it smooth. Size is often made from fine clay.

Very thin paper

*Books such as encyclopedias and dictionaries have hundreds of pages. They are printed on very thin paper. Thin paper can be strengthened by adding long **fibers** and glue to the **pulp.***

Paper for News

Newspapers are printed on paper called newsprint. Newsprint is thin, which means it is light but tears easily. Ink sticks to it very well, making it easy to print on. Newsprint often turns yellow if it is left in sunlight. Newsprint is also used to make magazines and phone books.

Millions of tons of newsprint are printed and **recycled** every day.

Newsprint must be strong enough not to tear as it runs through the printing press.

Newsprint is made from **mechanical pulp,** which is cheaper than **chemical pulp.** The **pulp** is a mixture of fresh pulp from trees and pulp from old newspapers. Newsprint comes in huge rolls. The rolls weigh up to a ton each and can contain 9 miles (15 kilometers) of paper.

Do not use it!

The surface of newsprint is too rough for printing fine details, so we do not use it for things like hardback books or photographs. We do not use more expensive paper to make newspapers because newspapers are usually thrown away or recycled after being read.

Folding and Cutting Paper

The **properties** of paper make it is good for making **disposable** objects such as party hats, paper bags, and wrapping paper. It is easy to cut and glue. Paper is also easy to bend. When you fold paper, the folds stay in place. So we can make things from paper by cutting, folding, and gluing. Objects such as envelopes and paper bags are made by machines. The machines cut shapes from paper, then fold and glue the paper into the correct shapes.

Paper bags must be strong enough to hold plenty of groceries without splitting.

This origami bird was made by folding a single sheet of paper.

Papier mâché

If paper gets soaked with water, it gets mushy. You can squeeze wet paper into a ball or mold it into any shape you want. The mushy paper is called papier mâché. When papier mâché dries, it gets hard and can be painted.

Origami is the art of folding paper to make models without cutting or gluing. Origami was invented in Japan. Special origami paper is thin and strong so it can be folded many times.

Tough Paper

Newsprint and writing paper are easy to tear. This does not matter because they do not need to be very strong. For other jobs we need paper to be tough.

Paper is made stronger by adding long cotton **fibers** to the **pulp** before the paper is made. The cotton fibers are stronger than wood fibers, and they hold together better. Paper can also be made stronger by adding glue to the pulp.

These bills from other countries will last more than a year.

Using sandpaper will make this wood feel smooth.

Sandpaper

We use sandpaper for smoothing wood and rubbing away old paint. It is made by gluing grains of sand to thick, tough paper. The sand makes the paper's surface rough.

We need very strong paper for some jobs. Brown wrapping paper contains cotton fibers that make it strong. Paper is made even tougher by adding very long, strong fibers from plants such as jute and flax. Paper made this way is very hard to tear. We use it to make tough, long-lasting paper objects such as paper money.

Paper Colors

Paper made from wood is naturally brown. Brown paper is used to make things that are **disposable** or do not need to be written on, such as paper bags and packing paper. White paper is made by **bleaching** the **pulp.** Bleach takes the brown color from the **fiber.** White **pigments** can be added to make the paper even whiter and brighter.

Brown paper is a cheap packaging **material.**

Brightly colored party streamers are made by cutting dyed paper.

Colored paper is made by **dyeing** or printing. A colored dye is added to bleached pulp before the paper is made. The color goes right through the paper. Colored paper is also made by printing colored inks onto the surface of white paper. This is how patterned paper such as wrapping paper is made.

Do not use it!

Paper made from ***mechanical pulp*** *contains other substances from wood besides fibers. These substances gradually turn yellow if they are left in sunlight. We do not use this type of paper for expensive books. This is because we want the pages to stay white for many years.*

Paper Decoration

Colored and patterned paper is used for all kinds of decorations. Wallpaper is thick paper with patterns or pictures printed on it. It is glued to walls for decoration and color. Most wallpaper is smooth, but some is **textured.** **Vinyl** wallpapers have a thin layer of plastic that stops water from getting into the paper. This makes the paper easy to clean.

Hanging wallpaper is a quick way to change the color of a room.

These Mexican piñatas are made by curling many strips of colored paper.

Do not use it!

Paper catches fire easily and burns well. So we do not use paper to make things that get very hot, such as cooking pans. The paper would burst into flames. We must use ***materials*** *that withstand heat instead, such as metals or* ***ceramics.***

Pieces of paper are easily glued together or cut with scissors. A paper collage is a picture made from paper shapes that have been glued onto a background. Paper is also used to make hanging decorations for parties.

Cardboard

Cardboard is like very thick paper. It is made by gluing several sheets of paper together. The **fibers** for cardboard come from wood, wastepaper, and straw. Cardboard is a cheap, strong **material.** We use it for book covers, folders, and the backs of pads of paper. Cardboard is also made into tubes. Tubes are used for storing and protecting rolls of paper and large sheets of paper such as posters.

The cardboard tube will protect this poster from damage.

When cardboard needs to be even stronger, a sheet of thick brown paper is bent up and down to make a wavy pattern. Sheets of flat cardboard are then glued to each side of this sheet.

Do not use it!

Thick cardboard cannot bend without breaking. This means we cannot fold thick cardboard into envelopes or bags. Cardboard is also too thick for making newspapers and most books.

This type of cardboard is very strong. It is called corrugated fiberboard.

The wave-shaped paper gives corrugated fiberboard extra strength.

Boxes and Packaging

Like paper, cardboard has many uses. One of its main uses is for packaging things we buy from stores, such as toys and food. Small items are mounted on pieces of cardboard with staples or plastic sheets. Cardboard is also folded into boxes. Cardboard sheets and boxes protect things from being damaged accidentally.

Because it is cheap, cardboard is used to package all kinds of objects.

Cardboard has protected this pizza and kept it warm.

A sheet of cardboard can be glued and stapled to make a box. Cardboard boxes are used to protect objects that are heavy or easily broken such as televisions. They also protect things like books and pizzas as they are carried from place to place.

Waterproof Paper

When paper is made, the **fibers** stick together naturally. When paper is soaked in water, the fibers break apart again. So paper quickly becomes weak when it gets wet. Eventually the paper falls apart. Even thick, strong cardboard falls apart when it gets wet.

A wet corrugated fiberboard box loses its strength very quickly.

Wax on the inside of this juice carton keeps it from falling apart.

Paper and cardboard can still be used for jobs in which they get wet. Milk and juice cartons are made from cardboard. One way to make paper **waterproof** is to rub wax into the surface. Wax **repels** water, so water cannot soak into the paper. Another way is to add a very thin layer of plastic to the paper. Plastic is waterproof, so it keeps water from getting into the paper.

Do not use it!

Most writing and printing inks have water in them. They work by sticking to the surface of paper. If we tried to use inks on waterproof paper, they would not dry well and would smudge. So we do not use waterproof paper for writing.

Tissue Paper

Paper is good at soaking up water. We say that it is **absorbent.** Tissue paper has many tiny spaces between the **fibers** that soak up water. We use tissue paper to make paper tissues, paper napkins, toilet paper, and paper towels.

Some tissue paper, such as paper towels, must be strong when it is wet so it does not fall apart. We make tissue paper stronger by adding **resins** to the **pulp** before the paper is made. The resins stick the fibers together.

Paper towels can soak up most liquids.

Paper masks are used in hospitals to protect doctors and nurses from germs.

Tissue paper is used to make air filters for vacuum cleaners and masks. The tiny gaps between the fibers let air pass through but not bits of dust or soot.

Do not use it!

Most tissue paper is very weak. So we do not use tissue paper when we need strong paper. For example, we do not use tissue paper for making paper bags.

Paper and the Environment

We throw away many tons of paper each day. Paper rots away naturally, so you might think throwing paper away is not a problem for the **environment.** Paper mills use a lot of **energy** to make paper, however. This energy can be saved by **recycling** paper. Recycling also reduces the number of trees we cut down. It limits the amounts of **chemicals** used to make **pulp** and to **bleach** paper. Most newsprint and cardboard contain recycled paper.

Recycling bins like this one can be found in most towns.

Newspapers can be a good bedding **material** for some pet cages.

Paper recyclers collect wastepaper and cardboard packaging from recycling bins in factories, offices, and stores. You can help, too, by taking your wastepaper to a recycling center.

Do not use it!

Recycled paper is used for cardboard. It is hard to remove all the ink and other chemicals from wastepaper, though. Most recycled paper is not white and bright like paper made from new pulp, so we do not use it for making expensive hardback books.

Find Out for Yourself

The best way to find out more about paper is to investigate it for yourself. Look around your home for things made from paper, and keep an eye out for paper during your day. Think about why paper was used for each item. What properties make it the best material to use? You will find the answers to many of your questions in this book. You can also look in other books and on the Internet.

Books to read

Balchin, Judy. *Papier Mâché.* Chicago: Heinemann Library, 2000.

Ballard, Carol. *Science Answers: Grouping Materials.* Chicago: Heinemann Library, 2004.

Donald, Rhonda Lucas. *Recycling.* Danbury, Conn.: Scholastic Library, 2001.

Hunter, Rebecca. *Discovering Science: Matter.* Chicago: Raintree, 2003.

James, Diane. *Paper.* Chanhassen, Minn.: Creative Publishing, 2004.

Stevens, Clive. *Origami.* Chicago: Heinemann Library, 2002.

Watson, David. *Papermaking.* Chicago: Heinemann Library, 2003.

Using the Internet

Explore the Internet to find out more about paper. Have an adult help you use a search engine. Type in keywords such as *paper mill, cardboard packaging,* and *paper recycling.*

Glossary

absorbent able to soak up liquid

bleach chemical that removes the color from materials

blotchy having colored patches or stains

ceramics something made from clay that has been heated until it gets hard

chemical substance that we use to make other substances, or for jobs such as cleaning

chemical pulp pulp made by breaking up wood with special chemicals

disposable can be thrown away after it is used

dye liquid added to something to change its color

energy power to do work

environment world around us

fiber long, thin, bendable piece of material

material matter that things are made from

mechanical pulp pulp made by crushing wood

mesh metal net used to separate a solid from a liquid

microscope instrument used for making things look bigger

natural anything that is not made by people

pigment substance added to paper to give it color

property characteristic or quality of a material

pulp mixture of fibers and water that paper is made from

rag pulp pulp used together with chemical pulp to make high-quality paper

recycle use again

repel force back or keep away

resin thick, sticky sap made by certain plants

textured having a rough surface

vinyl type of plastic

waterproof word that describes a material that does not let water pass through it

Index